D0878986

The PLM Primer

Stephen Porter

Steve Porter, CEO, Zero Wait-State
6207 Bee Cave Rd, Suite 260, Austin, TX 78746
Phone: 512-732-0721 x302
            Fax: 512-682-9099
            Cell: 512-921-2912
sporter@zerowait-state.com
www.zerowait-state.com

ISBN 978-0-9886183-0-5

10 9 8 7 6 5 4 3 2 1

Contents

Introduction

This book is a compilation of three years of blogs about Product Lifecycle Management (PLM). In my role as CEO of Zero Wait-State my goal has been to provide the PLM industry with useful information to help companies make the right choices about improving their product development process. This book is meant to provide practical unbiased advice on how best to move forward for organizations looking at PLM or ones in the process of implementing or leveraging this technology.

The knowledge shared in this book is equal parts experience and research. I have spent the last ten years helping to build Zero Wait-State into a first class organization that offers high level consulting for PLM. During that time I have had the opportunity to work alongside some very bright individuals both in consulting and in companies leveraging PLM technology. I appreciate these compatriots sharing the benefits

v

of their knowledge and experience with me, and am happy to pass it along to the readers of this book.

On the research front I have been closely following several authors who have contributed quite a bit of material to the PLM mindspace, especially the prolific Oleg Shilovitsky and the thoughtful Jos Voskul. Reading their thoughts and opinions on PLM has inspired ideas on numerous occasions, and I certainly have appreciated their feedback on my articles. I must also thank authors Seth Godin, Daniel Pink, Malcolm Gladwell, and Jim Collins for writing very thought provoking books that can be applied to companies striving to improve their product development process.

I hope you find *The PLM Primer* to be a useful resource in your journey toward optimizing your product development environment. I look forward to adding more material to this book as I write more articles and leverage material from my coworkers at Zero Wait-State and elsewhere.

CHAPTER 1

Why PLM?

How Fast Can You Turn Your Ship?

Companies and ships have a lot in common, particularly when it comes to navigation. As the sinking of the *Titanic* illustrated, ships are not the most maneuverable vessels and companies share this trait. Trying to respond to major challenges or changes in your company can be difficult, especially if information is scattered throughout the company in multiple systems that manage business processes. Product Lifecycle Management (PLM) helps companies organize their product data and controls processes to allow for more dynamic responses to the demands customers and market conditions make on organizations.

Jim Collins' latest book *Great By Choice* (written with Morten T. Hansen) refutes the idea that innovation alone is a guarantee for survival in turbulent times. Collins touts companies with fanatical discipline, empirical creativity, and productive paranoia

1

as the ones able to weather the rough seas of corporate life. This chapter will illustrate how PLM enables companies to respond quickly to make the needed adjustments for survival and success, and we'll see how PLM promotes the virtues Collin's touts.

In a chapter titled, "Fire Bullets Then Cannonballs," Collins discusses the danger of an all-out innovation approach. He cites another book titled *Will and Vision* by Gerald Tellis and Peter Golder that researched 66 product market segments and found that only 9 percent of innovators end up as the market leader. Their research shows only three of the seven higher performing companies had a better track record in innovation versus their comparison company. Collins' conclusion is that, "The great task, rarely achieved is to blend creative intensity with relentless discipline so as to amplify the creativity rather than destroy it. When you marry operating excellence with innovation you multiply the value of your creativity."

This "operating excellence" is where PLM comes into play. Having your product record captured and fully defined in a single system gives a company enormous control over the product development process. Controlling all of the information that defines a product and having the ability to capture the process allows companies to measure and quantify the time it takes to develop these products. If you do not have PLM in place doing this can be almost impossible unless you are an extremely small organization or developing very simple products.

There are a number of company objectives that make PLM necessary especially if one aspires to maintain operational excellence. Last year Brian Shepherd from PTC listed several in an article titled, "Break Down the Barriers to Product Development" published on the Cadalyst site. He cites the need for distributed design, extended collaboration, secure data sharing, complex program management, and managing change throughout the lifecycle as key business drivers for PLM. Most of these are things companies encounter as they grow. Having PLM in place before you encounter these issues provides a huge business advantage and allows more effective utilization of these approaches to expanding product development capabilities.

Think about the logistics involved in managing distributed design without a PLM in place. How do you ensure that the remote groups, like design teams or contract resources have the correct information? How much time will you have to spend making sure the work is done accurately? How do you take this information and pull it into the internal product record for manufacturing, purchasing, etc.? Change management is another huge problem without PLM. How do you make sure you are making the right versions of products or that engineers aren't building new versions on top of old data? Not having PLM makes it very difficult to accomplish types of activities well and slows everything down significantly. With PLM in place maneuvering through these challenges becomes much easier and the helm of your boat is much more responsive.

Some individuals and companies view PLM as overhead that interferes with product development, another level of bureaucracy that gets in the way, a redundant system, or less important resource than Enterprise Resource Planning (ERP). Without PLM most companies will struggle to respond to industry and customer needs. They will make decisions without critical information and fire blindly to try and hit their customer's requirements. PLM allows a company to organize their data and processes for product development which puts them in the best position to mine information for good decisions and identify bottlenecks in their product development process.

Innovation by itself is just a fast way to spend money. It must be coupled with operational discipline that ensures efforts are targeted and efficient. By following this formula it is possible that your company could be cited by Collins in his next book. Being able to design for cost and compress time to market can be the difference between succes and failure. At the very least it will certainly allow you to steer clear of icebergs.

CHAPTER 2

What Can PLM Do for My Company?

Go With the Flow: Using PLM to Change and Inspire Your Company

Flow, according to poet W.H. Auden, "is the state of forgetting oneself in a function." Whether its riding a wave or hitting a golf ball the feeling of "being in the flow" is unsurpassed. Encouraging change or adoption of new tools is one of the biggest challenges facing companies and vendors today. Tapping into the emotion around "flow" can be a powerful motivator. Change can best be attained through creative means. When attempting to justify Product Lifecycle Management (PLM) solutions, like AgilePLM from Oracle or Windchill PDMlink from Parametic Technology Corporation, dry numbers on return on investment don't create the kind of emotion you need to trigger change.

An often overlooked methodology when attempting to either justify or implement process improvement programs like PLM or engineering collaboration is empowering the masses. The power of the people has been essential to drive major change in cultures and in industry, and yet it has been underestimated throughout history. My definition of "the masses" is the people at the sharp end of the industrial stick: engineers, shop floor resources, and configuration management resources. These people will be most impacted on a day to day basis with process change. If you don't have their buy-in and input from day one the likelihood that you will succeed with a major process change initiative is highly unlikely. Moreover, if you can leverage these people as resources and get them behind the project, PLM solutions can have a transformative effect on their attitudes toward their jobs. This effect on their attitudes can yield far more benefit to a corporation from productivity gains than the technology itself.

According to a Gallup poll cited in *Drive: The Surprising Truth Behind What Motivates Us* by New York Times bestselling author Daniel Pink, nearly 20 percent of the U.S. workforce is actively disengaged at work, costing companies nearly $300 billion dollars a year in lost productivity. By utilizing a strategic approach that actively engages all personnel impacted by process initiatives, companies can dramatically increase a program's effectiveness and boost productivity overall.

I realize this is easier said than done and that "too many cooks can spoil the broth," but there are definite ways to solicit input and to engage with people at all levels of a company to create a groundswell of support for process improvement initiatives. By framing PLM as a solution that will remove the tedium from their jobs and empower them to reach their full potential, organizations can dramatically affect the momentum for change and the results of said change.

This chapter will explore some of the ways PLM and engineering collaboration can have a transformative effect on day to day tasks and engage employees, who will increase their productivity through better motivation and more efficient process and tools.

There are numerous business philosophies and strategies about how to empower and motivate employees. Most of these philosophies look at technology as a necessary evil or as a means to an end not as an end itself. I propose that PLM is a means to empowerment and motivation if properly implemented. Proper implementation means allowing all levels to participate in selection and implementation of the tool. I have been involved with many PLM projects where IT or management exclusively selected the PLM solution and how it would be deployed. In most instances acceptance of the PLM across the company was problematic, and if management disengaged then the project floundered without them there to enforce adoption.

How you present PLM to the company is important. If it is perceived as another level of bureaucracy that someone has to wade through to get their job done then it will be problematic. If employees see this as a way to automate the repetitive tasks in their lives and eliminate the need to recreate information then I suspect their attitude about the solution would be different.

All of this points to the way a PLM solution is selected. Committees have a bad reputation for dragging things out and complicating issues. One only has to look at Congress to know this is true. However there are ways to solicit input and to promote ideas without burdening a project with the complexity of a committee. Well-thought-out user surveys and transparency throughout the selection process can go a long way toward winning support for process improvement. As long as employees are kept informed about the selection process and the reasons driving the selection most will be enthusiastic and supportive. Several companies we have worked with have leveraged up front user surveys to great effect. This allows the end user to feel that he has had some input into the decision making process. Another great technique is to sponsor "lunch and learn" events featuring some of the technologies being considered. Another method is to hold town hall style meetings to discuss issues in the company and what is being done to address these issues.

The key to engaging employees in an initiative like PLM is to enable them to achieve "flow" with the new tool. In *Drive,* Pink cites research from several clinical psychologists indicating that the most satisfying experiences in people's lives were when they were in the "flow." To be in the flow, goals must be clear and feedback must be immediate. The relationship between what a person has to do and can do is perfect. Properly implemented and adopted PLM can help people achieve "flow" as Pink describes: "In flow people lived so deeply in the moment and felt so utterly in control, that their sense of time, place, and even self melted away. They were autonomous of course. But more than that, they were engaged."

Anyone who has worked with a properly configured PLM solution can understand the relationship between the state described here and their experience. What could be more empowering than having information at your fingertips? Having the ability to see at what stage a change order is in your process. Being able to work freely and not be dependent on the availability or disposition of a coworker or supplier. Not having to interrupt a design task to satisfy the query of a co-worker. The key to helping employees see PLM in this light is to solicit their input early in the process and to ensure that they are fully trained with the tool. One of the biggest issues we see from an implementation perspective is companies that try to save money by reducing the number of people trained or the amount of training delivered. Selecting PLM solutions that have more intuitive interfaces can also affect this dynamic in a positive way.

The takeaways from this discussion are that PLM, when properly selected and implemented, can deliver value far beyond the specific reductions in time that automation provides. It can create a higher level of satisfaction and empowerment in the employees of a company that will affect their motivation, productivity, and overall attitude. This morale affect will yield benefits far beyond what PLM can deliver. The key is to engage with all levels early on and solicit input and incorporate this input in the selection and adoption methodology. On the flip side, a poorly implemented PLM that is mandated from above without the input of the organization will not only fail but can create dissention and dissatisfaction that will further damage a company by decreasing productivity through alienating and demoralizing the employees.

In order to be successful one must "go with the flow." Involve all levels and ensure that everyone is fully trained to achieve their goals. Engineering an environment for success amplifies the value of PLM. Vendors who recognize this approach will become valued partners to their clients, and companies who insist on this type of approach will ensure strong return on investment. As one study cited in Pink's book concluded, "... the desire for intellectual challenge, that is the desire to master something new and engaging, was the best predictor of productivity." Theresa Amble, a Harvard professor quoted in the book adds, "The desire to do something because you find it deeply satisfying and personally challenging inspires the highest levels of creativity whether it is in the arts, sciences, or business."

PLM can enable this type of environment for a company by eliminating drudgery and empowering employees to control their own destinies. Vendors and companies must make the extra effort to carry this message and equip their employees with the proper tools and training to allow them to achieve a state of "flow."

CHAPTER 3

How Do I Justify an Investment in PLM?

If the Glove Doesn't Fit You Must Acquit

Johnny Cochran's rhyme popped into my head as I was reading a story from a great book titled, *Switch: How to Change Things When Change is Hard,* by Chip and Dan Heath. In the story a man named Joe Stegner is struggling with ways to convince his company, a large manufacturer, that they are wasting large sums of money. Sound familiar?

In this specific example he observes his company wasting billions of dollars through inefficient purchase processes. A major shift in process must occur to eliminate this waste, and he needs a compelling example to drive his point home. He seized upon a single item that illustrated his point: gloves. After researching the issue, he discovered that the factories in his company were purchasing 424 different types of gloves. Moreover they were purchasing gloves from various suppliers and costs could range from $5 at one factory to $17 in another. When it came time for his meeting with the executives at his company he piled a table high with all 424 types of gloves and the price paid for each pair. He titled this display the "Glove Shrine."

The executives were stunned. They could walk around the table and see different pairs of gloves that were basically very similar yet were costing the company $10 or $15 dollars more than others. The exhibit was so effective the company took it on the road so that everyone in each factory could see how much money the company was wasting just on gloves. In the end the exhibit had the desired effect. It illustrated the point that the company was wasting money through a flawed process, which motivated the company to address this issue. Everyone ended up winning except for the glove salesman.

The similarities between this story and the discussions we have with management to facilitate change via PLM and engineering collaboration are striking. Most of the time we present data via spreadsheets and Powerpoint and our audience nods their heads in agreement or just nods off period. We go through elaborate exercises to convince management of the need for change and often these presentations are ignored. The question becomes: how do we as agents of change identify and effectively communicate the impact of PLM to management at product development organizations?

The challenge of PLM is that the value isn't nearly as obvious as solutions to other problems. The glove example is pretty straightforward. Financial and manufacturing systems are more directly aligned with cost drivers, so financial justifications are easier to create in order to justify investment in these types of systems. PLM leads to secondary effects that create value. PLM doesn't develop products faster. It enables process optimization and data reuse, reduces late stage change, and brings manufacturing into the process earlier. All of these things ultimately allow a company to develop more products in a shorter period of time, but how much more and how much shorter really depends on how efficient the company was in the first place and the complexity of the products. The net result is that PLM return on investment presentations are very abstract and very dry.

Some management teams just get it and know intuitively that it makes sense to have a single source of truth with all physical data vaulted as a product record. They understand that electronic processes can be monitored and tweaked for maximum efficiency. They understand that having all of this information captured in a database can yield real time reports that will allow them to identify the bottlenecks in their process and maximize their efficiency. Studies by Aberdeen, CIM Data, and others bear out that best in class companies use PLM and benefit from it. Aberdeen's study, "Profitable Product Development for SME" states, "That Best in Class [small and medium enterprises] are meeting the product development goals that drive product profitability 49 percent to 74 percent more frequently than average companies, and up to 14 times more frequently than laggards." This is pretty compelling stuff, but how do you present it in a way that resonates for a specific company?

Joe Fowler published a blog on Linkedin that offers some vital perspective on the value of PLM. In his article titled "Critical Success Factors for an Enterprise Wide PLM," he writes, "Remember PLM is bigger than ERP. You must make the commitment. PLM is to your intellectual property what ERP is to your physical property. I would challenge and ask each of you which one is more valuable to a company." Fowler is in a position to speak authoritatively on the value of PLM. He was a leading resource at one of the biggest PLM initiatives in the history of the industry at Lockheed Martin and has spent a good portion of his career involved in implementing PLM solutions from the industry side.

So I think we can safely say that PLM is something that product development companies should have high on their list. Yet there is still significant resistance in the industry and many companies balk at the high price tag of enterprise-wide PLM.

To understand the reasons companies and their leadership teams resist the change required by PLM, you must understand psychology. Many of you are aware that psychologists theorize that the brain has two independent systems: an emotional side and a rational side. The book *Switch: How to Change When*

Change is Hard by Chip and Dan Heath dubs the two sides the "planner" and the "doer." In order to trigger action you must appeal to the emotional system rather than the rational one. The emotional system in our brain is far more powerful and often sabotages us when our rational side knows better. Think of things like diet or exercise where we know we should eat right and exercise regularly, but when it comes down to it, we often don't do it. Process improvement is not that different. Most executives know that it needs to be done but inertia, fear of the unknown, or just simple procrastination keeps them from acting. Powerpoint slides and spreadsheets are not going to appeal to the emotional side and motivate someone to action.

You have to find a compelling way to illustrate change, something that will reach people on an emotional level. Taking a list of companies that are directly competitive to your company and showing how they have adopted PLM and put your company behind would definitely trigger some emotion. Stacking the monetary savings with monopoly money might be a little gimmicky, but may still be more effective than an ROI spreadsheet. If you are passionate about improving your company's ability to effectively develop products you need to tap into emotion and use it to motivate others.

In summary, PLM presents a complex but signifi-cant value proposition for companies. While it can be difficult to quantify and communicate the value, your biggest opponent is human nature and the aversion to change. You must be aware of the psychological elements at play when you are advocating a departure from the norm and find unique ways to appeal to people's emotions in order to successfully revamp the status quo. It really doesn't matter if it's PLM or any other major process improvement. People will resist, managers will hesitate and ultimately the decision will be made based on emotion. You are competing with many other process improvements and capital expenditures so it is critical that you identify the elements of PLM that will trigger the "doer" rather than the "planner." Remember the glove, it worked for Johnny Cochran and it can work for you.

CHAPTER 4

Where Do I Start with PLM? Aligning IT with Management

Closing the Communication Gap between IT and the Executive Team

On September 8, 1974, Evel Knievel attempted to jump the Snake River Canyon in a steam powered Skycycle X-2. Between the parachute deploying early and the unanticipated headwinds, the attempt failed miserably. The Skycycle failed to even reach the water below, which was probably a good thing for Evel since I am not sure the rocket would float or that he could swim.

Watching IT and executives interact around technology and process initiatives often reminds me of this futile attempt at defying gravity. The gap between these two groups is often as wide as the Snake River Canyon, and often IT seems as ill-equipped as Evel was that day when it comes to communicating value to the executive team. These issues can also apply to other personnel within a company that are attempting to justify improvement initiatives to management. The main barrier to communication between executives and the rest of their organization is that they are looking at these projects from a different perspective and often know or care very little about the technology itself. Their focus tends to be financially oriented or tied to corporate objectives that are not IT related.

This chapter will discuss how to blend the interests of IT and engineering with the executive teams' objectives to facilitate adoption of beneficial technology and process solutions.

Recently I was involved in a discussion where an IT resource was trying to convey the value of virtualization to a manager. The IT person was very knowledgeable about the technology but failed to grasp what key pieces of information were important to the manager. While he dwelled on the capabilities the company would be gain by adopting this technology, the manager was more concerned about the cost of acquiring this technology and what resource level was going to be needed to sustain it.

While not every IT or engineering resource is oblivious to management's perspective it's worthwhile to discuss the idea of framing process improvement initiatives in a way that resonates with upper level management. Management is cost-driven, while IT and engineering are more concerned about capabilities.

This is about understanding the perspective of management and engaging in a dialogue and process that will result in mutual satisfaction for both engineers and managers.

It can be difficult to quantify value, both monetary and otherwise. It is very important to have this information ready when attempting to promote a process improvement initiative. I was recently scanning back through an excellent book on this topic called *Let's Get Real or Let's Not Play: Transforming the Buyer/ Seller Relationship* by Mahan Khalsa and Randy Illig. I consider this book a must read for any salesperson or consultant involved in process improvement technologies. I also think that those who are on the corporate/ technology adopter side could also benefit from this book. It gives great insight into the pressures applied to both sellers and buyers and how to ensure a positive outcome when purchasing process improvement technology. Kahalsa and Illig write about the thoughts a manager might have when presented with a new idea. If the manager is smart he might ask the question, "Before we write a check for this, could someone please tell me how this is money well spent?" This seems pretty obvious, but I have been involved in many projects where this question was not asked at all or was asked very late in the project.

While quantifying the financial value of a technology investment is a best practice for a sales or consulting company to promote, many organizations view this as extra work and don't really push it unless the customer insists or the sales cycle stalls. It is incumbent on the client to drive this activity, and most likely a manager will raise this issue so it is important for engineering and IT to be ready with strong justification to address this question.

When trying to justify PLM initiatives to management, talking about the functionality that will be gained is probably not going to resonate. Notice the question, "How is this money well spent?" It could be argued that gaining certain capabilities justifies the expenditure, but if you make this argument you need to go a step further and quantify the impact of having or not having these capabilities.

In the example I used above about virtualization it might be that not having a virtualized environment for your PLM system means you must spend money on extra hardware for test systems and that the backup only covers the data stored in the system but not the system itself. That alone is a pretty good argument, but you could take it further by explaining that it takes five hours a week to perform and verify backups and maintain multiple hardware platforms and that over a twelve month period using a burden rate of $100 per hour this adds up to over $20,000 in extra money spent. This doesn't even factor in the cost of the extra hardware and the time lost if the system has to be reinstalled.

Managers can understand this type of justification. You are spending extra money on IT resources (this is especially compelling if the IT resources are contract) and hardware, and backup problems expose the company to further risks. Vendors like VMWare and Oracle actually have spreadsheets for calculating cost and calculating value, but these tools tend to be somewhat generic and are viewed with skepticism by management. A management justification for this type of initiative needs to be based on cost savings that are closely tailored to your environment and that will resonate with your organization.

There is a good set of questions IT professionals and engineers should ask themselves before even thinking about approaching management with an idea. This is straight from Khalsa and Ellig's book.

- How do you know there is a problem?

- What lets you know there is a problem?

- Where specifically does the problem show up?

- Which measures prove there is a problem?

- Who specifically is most affected by the problem?

- When does this problem occur most?

These questions establish evidence that there is a problem, but you need to take it a step further and determine if this is a problem worth solving. These questions address quantifiable results.

- How specifically would you measure success?

- What would let you know you were successful?

- Where would the success of this project show up?

- Which performance indicators will increase or decrease if we are successful?

- Who specifically would be affected by these issues, and how would they be affected?

- When do you need these results in place?

Some of these questions are redundant, but I wanted to keep them identical to the Khalsa and Ellig's book because sometimes if you ask the same question a different way you get a different answer.

The bottom line is to go through some significant analysis of the issue prior to approaching management. If you are management you should expect this level of analysis from your resources or vendors before you go too far down the road with a project. Evidence must be gathered prior to beginning a process improvement project or its value may be lost in the changes required to make it work after the fact. Once you have gathered this information you need to quantify the impact like I did in the virtualization example. Khalsa and Ellig's book goes into methods for doing this, but keep in mind that there are no magic wands for this activity.

If Evel had stopped and taken the time to better analyze his attempt to jump the Snake River Canyon, we might have been deprived of the spectacle of him plummeting to the ground in the red, white and blue Skycycle. I suspect his main calculations that day were for the likelihood of his survival and in that regard everything worked out. Before you decide to jump off a canyon like purchasing PLM or adding functionality, ask a few questions and come up with a plan that will ensure a safe landing on the other side.

CHAPTER 5

Which Type of PLM is Best?

Engineering-Driven PLM Versus Enterprise-Driven PLM

My son is mesmerized by Rick Riordan's *Percy Jackson & the Olympians: The Lightning Thief.* In the book the gods of ancient Greece battle with their unique abilities, which results in epic conflict. Currently in the PLM space there seem to be two types of PLM solutions with unique abilities that vie for supremacy. Several solutions, including Windchill PDMLink from PTC, Enovia from Dassault, and Teamcenter from UG/Siemens, have their origins in computer aided design (CAD) and derive a lot of their structure directly from engineering data. Alternatively, products like AgilePLM from Oracle, Aras, and Arena are driven by the bill of material and seem to be more focused on the manufacturing side of organizations.

All of these products fall under the same label and presumably address the same business challenges but do so in a significantly different way. It presents an interesting dichotomy. Whether you choose lightning bolts or tridents, slaying inefficiencies in product development processes is no trivial feat. Potential customers need to fully understand the benefits and disadvantages each type of PLM presents so they can determine the best approach for their environment.

There are actually a lot of similarities in product development and Greek mythology. If you chop off one of the mythical Hydra's heads, two grow back; it's similar to solving one bottleneck or issue only to see additional problems created by the solution. I am sure we can all relate to poor Sisyphus who is doomed to forever push the boulder to the top of the hill only to have it roll back down to the bottom. It seems also that the conflicts between engineering and manufacturing are as ancient as the conflicts of the Greek gods and maybe even older. I can imagine the frustration of the first caveman drawing his wheel and seeing the disparity between the drawing and the finished product.

Zero Wait-State has worked with companies that deal with the complications of product development and has exposure to both types of PLM. Understandably, vendors for these products feel they are the superior solution in all cases regardless of the circumstances. However, we have identified that each approach has strengths and weaknesses that can be challenging for certain types of companies.

At first glance most PLM solutions seem remarkably similar. They all are based on a database (usually

Oracle). They typically have a web-based user interface with browser friendly conventions. They all have a pretty robust access control logic that allows for granular control of what information can be accessed by whom. Finally, some sort of rules-based workflow engine is essential for any PLM. When companies evaluate PLM they compare these features and tend to settle on the one whose capabilities seem to best fit their needs. Some PLM solutions have more granular security. Some have a more robust workflow engine. Others have a more intuitive user interface. Sometimes when the PLM vendors fail to differentiate themselves from the competition, the decision is driven by price.

However, one area engineering-based PLM vendors focus on more heavily is CAD data management. CAD data presents some unique challenges. It is typically hierarchal and requires interfaces from the CAD system itself to correctly map the relationships into the PLM. Some vendors don't expose these interfaces, giving them an advantage in managing this information in their specific PLM solution.

On the other side, enterprise-based PLM solutions emphasize enterprise connectivity issues, integration into Customer Resource Management (CRM), and Enterprise Resource Management (ERP). They will imply that having all of these applications on a unified framework is advantageous, and sometimes it is, depending upon the types of products developed and the opportunities for cooperation between engineering and other organizations within a company.

This is where the tradeoffs begin. Storing CAD data inside a PLM requires certain types of data models be in place and takes up a significant amount of bandwidth due to the size and complexity of these files. It also dictates that the interface be significantly more complex. On the other hand not having this information vaulted in the PLM can create a disconnect between engineering and manufacturing, which is one of the main things you are trying to avoid when you adopt PLM. It comes down to speed versus power in some cases, and it really depends on the type of company you have and the challenges you are facing. Certain types of companies structure their design closely around their engineering data. These companies benefit from a closely aligned system because downstream information is derived from the engineering data. Other companies will dramatically alter the engineering data due to manufacturing requirements. These companeis see less value from having CAD data fully integrated into PLM. The key is to understand your company's requirements and choose a PLM that best matches your company's needs.

Each of the gods in the Greek pantheon had unique powers. Zeus had his thunderbolts and could blast his foes into oblivion, but sometimes he wasn't particularly precise in doling out his wrath. Hermes was the messenger of the gods: he had wings on his feet and could move about with unrivaled speed. Athena was the goddess of wisdom: she knew all things and used this knowledge to her advantage. Each of these gods is similar to the various PLM solutions with their positive capabilities and the downsides that come with them. Power means complexity; ease of use means potential limits on capability; customization means an implementation nightmare. When choosing the best fit for your company, recognize that compromises may be necessary to serve the greater good of the organization. The challenges of CAD data management can force you to choose an engineering-based tool, but the complexity may be too burdensome for the rest of the company. If you understand where the true value of PLM resides, you can better understand the path you should choose.

CHAPTER 6

How Do You Bridge the Gap to Engineering?

Engineering Collaboration: No Organization Is an Island

Historically, engineering groups tend to keep to themselves and view outside communication as a necessary evil and an inconvenience. This communication might be in the form of a drawing or even a CAD model or viewable. It involved a manual process where the information was transferred from the native environment of engineering via a network drive, email, FTP, or even paper. Terms like "silos of information," "black box," and "Bermuda triangle of product development" are often used to describe the engineering department. One of the more prevalent terms used is "Island of Automation."

Most engineers tend to embrace technology and are early adopters of solutions that help them work more efficiently. So often (not always) the engineering organization might be ahead of the rest of the company when it comes to having their information organized and secured in a data management technology. In fact, tools like Pro/Engineer, UG, Catia, and SolidWorks almost require a system be in place to control access to information. Once you get beyond about five users, it becomes very difficult to control versions of files particularly if multiple people access the same data. So by necessity engineering groups adopted product data management solutions to address these issues.

Unfortunately, it was a very rare thing for an engineering group to expose these systems to anyone outside of engineering. Usually the data management tool itself was ill-suited for anyone outside of engineering anyway. Thus the "island of automation." The CAD vendors addressed the issue by making the island bigger by enhancing or replacing the CAD PDM applications with a more robust solution, and manufacturing oriented vendors responded by building bridges to the island via CAD integration. This chapter will contrast the approaches and the implications of each approach.

Most CAD vendors have their roots in CAD data management, which is why their PLM offerings typically excel at it. PTC and Siemens UG in particular have a long legacy of CAD data management. Even Dassault's Enovia/Matrix One product was developed originally to manage Cadra data. Of the primary PLM vendors, only Oracle's Agile has no connection to CAD.

PTC went through several different tools starting

with Pro/Project, then Pro/PDM, then Pro/Intralink, and now Intralink version 9, known as PDMLink. All the versions of data management prior to Intralink 8 and PDMLink were primarily focused on managing the files generated from Pro/E. PDMLink (Windchill) was actually a completely different application that PTC acquired with their purchase of Computervision, another CAD vendor. Jim Heppelmann, the current PTC CEO, developed the initial versions of the Windchill while he was employed at Computervison. Teamcenter was originally known as Metaphase and was part of the SDRC product line and was initially used to manage data from this system. Teamcenter has emerged from the mix of data management tools that have been cobbled together through Unigraphics acquisition of SDRC and then the subsequent acquisition by Siemens.

Oracle's AgilePLM was developed as a bill of material management tool and evolved into a fairly robust supply chain tool for the hi-tech and medical device industry. Along the way they acquired a German company called Eigner that had their own PLM solutions and some fairly sophisticated CAD integration applications. By acquiring Eigner, Agile hoped to bolster their capabilities in the CAD integration market and expand their European presence. They also acquired Cimmetry Corporation that produced AutoVue, which was a widely adopted CAD viewing tool to further enhance their functionality for interacting with engineering data. Oracle then acquired Agile to round out their enterprise application strategy as a complement to their CRM and ERP offerings.

The main take-away from this history lesson is that both PTC and Siemens products are deeply rooted in CAD data management and both companies have leveraged a strategy of building off of this legacy that has significantly impacted their approach to PLM and the types of solutions they offer.

As I pointed out in the introduction, both PTC and Siemens had created islands of automation. Their products could successfully manage CAD data, but they had to learn how to parlay this into an enterprise solution. It is difficult to see how current versions of the PLM offerings from both companies resemble the CAD-dedicated applications their customers originally utilized. In PTC's case there is a fairly clear break with Intralink 3.4 as the last version of Intralink that was based on the CAD-specific product that was offered several years ago as a replacement to Pro/PDM. When PTC introduced Pro/Intralink, customers were required to migrate off of Pro/PDM at significant expense and effort. Intralink, however, was a significant improvement over Pro/PDM for most companies and most of the companies using Pro/PDM were fairly large so the return on investment was usually fairly positive. With the introduction of Wildfire 5, current PTC customers are required to upgrade from Intralink 3.4 to Intralink 9, which is basically a slimmed down version of PTC's enterprise solution PDMLink. Customers are required to migrate and invest in more robust hardware and will likely need significant training to leverage the new version that really provides the same function as the previous version.

This has created some issues for PTC customers particularly some of the smaller companies. PTC wanted to reach out to the enterprise so they developed a new CAD data management platform that addresses more capabilities like change management, configuration management, and other PLM functions. Several of their large clients have adopted this PLM platform and moved off of Intralink 3X, but many of their clients feel trapped on the now-larger island. This product is very powerful and is probably the best choice for a company whose primary need is to manage Pro/Engineer CAD data and leverage that information throughout the product development process.

Unfortunately, if you are a Pro/Engineer customer and this product does not suit your environment, PTC makes it difficult to manage their data in other vendor's PLM products. Some of the feedback from smaller clients is that the expense of migration is problematic and the new version is overkill for their purposes. Some mid-size and larger companies already have PLM solutions in place, so they have issues with the redundancy between PTC's offerings and their current solutions.

Siemens has experienced similar issues as they try to transition their clients from legacy versions of SDRC onto their UG NX platform and consolidate the various forms of data management that have been deployed among their clients.

The bottom line is that both vendors have made their product more robust and complex, but this approach has created some issues among their existing customers. If you find yourself in these circumstances, the key is to look beyond the immediate need of CAD data management and determine the true best interest of your company. Options include moving to a different CAD system that does not try and lock you into a single vendor for data management, or consider compromising on data management for CAD in lieu of a enterprise PLM solution better suited for your company. The final option is to keep the CAD tool and integrate the data management tool with a better suited enterprise PLM system.

By contrast AgilePLM and Matrix have always chosen to address engineering via connectors. These applications typically allow companies to store the relational CAD data from products like Solidworks, Pro/Engineer, AutoCAD, Catia, and UG NX in their main vault. But unlike the CAD-based PLM tools, their data models didn't always accommodate the complex CAD data structures and struggled to support all the types of data created in the CAD tools. A more successful approach was to attempt to integrate from the PLM to the CAD data management tool supplied by the CAD vendor. This approach allowed engineering groups to have all of the functionality offered by the CAD vendor but also allow the rest of the company to use the best tool from their perspective.

The downside to this approach is that some of the

CAD vendors don't always appreciate having integrations built to their data management tools and any upgrades of either the PLM or the CAD environment can potentially impact the connector between the systems. In other words, the bridges that Matrix One and Agile build to the CAD tools have to be maintained and can potentially be severed by inclement weather.

Each approach has its pros and cons. By expanding the island to be more inclusive, PTC and Siemens have allowed clients to expand engineering's reach, but they have created redundant capabilities and burdened some companies with a level of complexity they do not need or want. By building bridges to engineering via direct and indirect connectors, Agile and Matrix One have allowed the rest of the company to gain access to engineering data, but the programs sometimes have trouble incorporating it into their environment and the link can be potentially tenuous depending upon the CAD vendor.

PLM customers must assess what is in their best interest by evalauting their true requirements and how to fully grasp the implications of these trade-offs through due diligence. User testing and extensive benchmarking are highly advisable before embarking forward with either approach. This is common sense, but hopefully the examples of Siemens and PTC show how important a well-studied choice is.

Product data is the lifeblood of companies, so it is critical that the pathways to engineering flow freely. Whether the strategy is to turn the island into a continent or build lanes of access, it is imperative that PLM be inclusive of engineering information to reach its full potential.

CHAPTER 7

How Can I Leverage PLM to Improve Process?

Is Using PLM to Design Your Business Process Putting the Cart Before the Horse?

Recently I was sitting in on a discussion between one of our consultants and a client. The consultant has years of experience with product portfolio management (PPM) and was advising our client on how to best adopt this capability in the context of their product lifecycle management (PLM) solution. The client had recently seen a demonstration of the PPM module and was feeling a bit overwhelmed. The issue the client had was that they currently had no process around project management that would easily translate to the centralized approach the PLM module required. The client was questioning how useful the module would be based on their current environment.

My expectation normally would be that the consultant would agree with the client that there was no point in purchasing the module until they had a more mature project management environment with a process that could support the application. However, he surprised me a bit when he insisted that adopting the module and then using it as a framework to build a new process would be the best approach. His point was that trying to build a process from scratch would be somewhat overwhelming without architecture to build on. He also felt that since ultimately this process would reside in the PLM why not start out that way instead of designing a process and then trying to make it work inside the PLM.

All of this seemed to make sense to the client, and it got me thinking about PLM as a whole and our past experiences trying to integrate companies' processes into PLM. Certainly process should be altered to leverage PLM capabilities, but can PLM be used as a tool to actually develop process? Is it more efficient to use PLM to design process or develop process independently of PLM? The answers to these questions can impact how PLM is implemented and the expectations companies have about PLM when they purchase it. This chapter will explore how using PLM as a tool to develop business process affects adoption methodology and ultimately the impact PLM has on a company's productivity.

One of the historic drawbacks for PLM is that it is not particularly flexible. Once you establish process and structure it is very difficult to go back and modify this structure. This is why consultants spend lots of time up front with the client mapping their product development process and making sure it is accurately captured within the system.

Some of the easiest (and I use the term in a relative way) implementations involve start-up companies that do not have established practices in place. These companies are often looking for guidance about best practices they can adopt that will allow them to fully leverage their new system. Some start-up companies put off purchasing PLM because they feel that they need to become better established and understand their process more thoroughly. Based on my experience companies that adopt PLM early tend to do better from a process automation perspective. They avoid building processes that don't map easily into the PLM tool. Having to go back and shoehorn a legacy process into a PLM is one of the more challenging aspects of implementation.

The point, however, is about existing companies and their adoption of PLM or specific aspects of PLM, like product portfolio management, costing, governance and compliance, and engineering collaboration. What if their processes for these areas don't exist or are so far removed from automation they cannot be leveraged? The implementation process for start-ups teaches us that using PLM to design process is a viable approach and that extending this methodology should be explored further. Individuals responsible for adoption of new technology should consider reaching out to vendors about pilot approaches as a means to fully define a solution prior to formal adoption. Vendors should consider allowing some degree of flexibility for software licensing when it comes to a true pilot. Companies should expect some external cost, but using the framework of the software as a means to structure a process independent from existing infrastructure could be a very powerful and efficient way to design new business solutions.

While researching this chapter I came across a white paper by Antti Saaksvuori titled "Building a PLM Concept." Saaksvuori literally wrote the book on PLM. His book is titled, appropriately enough, *Product Lifecycle Management.* One point the white paper makes that is relevant to this discussion is that "[c]ompanies seldom recognize the fact that the PLM maturity of the company is too low to launch a large scale PLM system project for the first time. There simply is not enough understanding of PLM and its possibilities, but also its impacts the current way [*sic*] of doing things. Usually the case also is that the processes and practices of a company are not mature enough to be utilized in PLM context." He is saying that, in many cases, companies' current processes are not compatible with a PLM system.

Putting the PLM system in place early and then using it to build process seems to make the most sense in these cases. Yet often I do not believe that is the expectation that PLM vendors set with their clients. I would go further and state that most companies could probably benefit from re-implementing their PLM tools based on the knowledge they have gained from the first pass. Obviously, vendors would prefer a more aggressive approach where companies fully commit to the PLM and then implement it. But by using the PLM tool as a process development solution, a company could implement first, then if they liked the solution they would purchase the needed licenses. Even if they decided to go with another PLM solution the process development work would have value.

Even in today's climate with PLM vendors touting out-of-the-box capability and rapid implementation methodology, it would be wise to consider flipping things around and using a PLM tool as a means to capture company process before fully committing. There is significant value to be derived from a smaller purchase of software and consulting to design your systems, and then use this knowledge to roll out to a wider audience. This approach allows companies to gain significantly more value from fully adopting a properly configured PLM solution. Process transcends specific PLM tools, so once a company transitions from manual process to PLM they can use their process with most of the different PLM solutions available. Virtualization makes it very easy to set up a PLM tool outside the context of the current environment. As a result prototyping has become far easier than it once was. Even companies looking to expand their PLM beyond its current scope can use this approach by setting up a development environment they can use to drive a new process.

There are several different perspectives for this approach. If you are a person inside a company that is struggling with where to begin with process improvements like PLM, this approach would begin with reaching out to a PLM vendor or systems integrator to quickly standup a test instance of a PLM system. That system could be leveraged to set up test processes for evaluation. Given the circumstances it is even possible the system you begin with may not be the system that is ultimately deployed. Remember this activity is about establishing PLM-compatible processes that you can leverage going forward. Inexpensive systems like Arena and Aras might make excellent test systems for this type of work. You should still engage with an external resource to provide guidance on how to develop these processes, and once you have them in place you can continue to tweak and experiment until you have solid methods to transfer to a more permanent solution. From a vendor or integrator's perspective you should welcome the opportunity to work with a company like this, provided they have a true plan to transition from the sandbox to a permanent solution. Ideally the vendor or integrator can make the initial setup fairly painless and cost effective.

None of the things I have discussed here are necessarily innovative. Many companies have broken up PLM into small segments to make it more manageable and shorten the ramp to value. Trying to undertake a long, complex, and expensive project to adopt PLM is not going to go smoothly. The unique idea from this chapter is that there is value in using PLM to construct new processes that are compatible with automation—outside of the typical value PLM delivers in automating process, shortening cycle time, and improving communication. The architecture and structure of PLM provides a platform that will be far more useful for developing process than trying to do it via committees with whiteboards and Powerpoint.

CHAPTER 8

How Do I Implement and Select a PLM System?

Plan for the Future

PLM solutions become an integral part of a company's infrastructure. It is not a simple task to move to another PLM once you have outgrown one PLM application. Try to think ahead three years or more and envision your company's size and business requirements from this perspective. Features and modules that seem unnecessary today can become critical requirements in the future and render a PLM application obsolete. Make sure you have the capability to manage all types of data, including CAD files, and the ability to connect with external applications like client relationship management (CRM) and enterprise resource planning (ERP) solutions.

It is also critical that the PLM solution be able to provide seamless and secure connectivity to external sites. Avoid pared down applications specifically for small companies. These applications are usually loss leaders designed to pull companies in, offering minimal functionality with expensive upgrade paths. The investment that vendors put into these solutions usually pales in comparison to the research and development dollars they put into flagship products, and this disparity can dilute the quality and sophistication of the company's products.

Identify Solutions that Can Be Implemented in Stages

PLM applications can address a broad spectrum of functions including data management, product quality, project management, governance and compliance, and product costing. It can be overwhelming to try and address it all at once. Identify critical issues in your company's product development process and prioritize based on the impact of the issue and how quickly it can be addressed with the PLM solution. PLM applications that are modular are much easier to implement in this manner. Some offer modules like program management that can stand alone and be implemented separately or in parallel with core applications.

The key is to be realistic and avoid prolonged implementation times. Long drawn out projects for deploying PLM can be prohibitively expensive and the longer an implementation project takes the more likely it will fail. As a general rule of thumb six week increments are ideal particularly for smaller companies.

Beware of Customization

While most PLM applications can be tailored to a company's specific business requirements there is a difference between configuration and customization. Customization typically involves writing code that must be maintained separately. The code can sometimes manipulate internal actions and typically involves external functions to supplement the application. It might use an external spreadsheet to organize data or pull information from an external source. Configuration is just applying settings within the application itself to create certain types of system behavior. An example of configuration might be the setting of user permissions or the stages in a workflow. Some vendors will try and address functionality shortcomings by indicating the application can be customized to address certain key functionality requirements like workflow processes or security. In some cases this is unavoidable because of complex business issues, but it should be the exception not the norm. A customized application will be difficult to support internally and with future PLM partners, particularly when new versions are released. Some applications offer automation capabilities that can be leveraged by companies themselves and do not create too much additional overhead. The key is to avoid

47

customization that would make you overly dependent on a vendor.

Usability Is Critical

PLM solutions often look very similar at first blush.
Most PLM applications use a web client that ties into
some sort of backend database. Given that browser
functionality provides a consistent look and feel, it
can be difficult to discern subtle differences between
applications. Talking to users with firsthand experience is the best way to identify the capabilities and
ease of use of a PLM application, but if this is not available, spending significant time on the interface is
critical. It is not unreasonable to have the vendor leverage company-supplied information to closely mimic
how the application would function in your company's
environment so that you can closely follow how the
application flows.

Difficult to use PLMs or those with limited functions can minimize adoption and impair productivity,
so usability is a key element in making a good decision
about a PLM application. One other key is to engage all
levels of users in the assessment. Typically, high level IT
or engineering resources are involved in the decision
making process and their familiarity with sophisticated software applications can skew the evaluation
of ease of use. Make sure that all areas of the business
participate in conference room pilots and user acceptance training.

Demand References

The best measure of determining the capability of a PLM solution is to look at comparable companies and the results they have achieved. Vendors tend to over-simplify aspects of utilizing and implementing PLM. Real world examples can help you better understand what is really in store. If a company is ready to recommend a solution and spend time helping you better understand the application, it speaks volumes about the quality of the product and the relationship that the company has with the vendor. It can also help you learn from the mistakes others have made.

The key in checking references is to prepare for the interview. If you allow the reference or the vendor to control the meeting they can gloss over key issues. It is critical to make sure your specific concerns are written down and presented prior to the meeting so the vendor can find the appropriate reference and you can benefit as much as possible from the meeting.

Measure Everything

The true value of PLM is the impact it has on processes throughout the company. It is difficult to quantify this impact unless you have a good understanding of your current process and the time and effort it takes to complete tasks. Before you embark on purchasing a PLM system, engage with internal or external resources to document your current product development process and identify bottlenecks and areas that are hindering your company's ability to deliver quality products to market at expected margins.

A successful PLM implementation should allow a company to increase the number of products they deliver, minimize the amount of late stage change, improve design manufacturability and cost, and raise quality. If you do not have a baseline established, it will be impossible to determine if any improvements are being made or how substantial the gains are. Management will be skeptical about investing further in a solution that does not provide tangible value.

Once you have established baselines and priorities, you should require your vendor to build metrics gathering into the implementation process. It will benefit both parties. By attaining promised return on investment, the vendor will be able to validate their solution with you and other companies in the future, and you will be able to justify further investments in process improvement. It will also allow you to ensure the application is properly implemented and address the issues you have determined that are most critical for your company's success.

Don't Procrastinate

Companies are often tempted to put off infrastructure improvements like PLM if they're small or finances are tight. They find ways to get by with various manual and semi-automated solutions to develop and manufacture products. The best in class companies in any industry do not take this approach. They leverage any and all means to become as productive and efficient as they can. PLM is a crucial building block to any product development company's foundation. PLM takes the guesswork out of product development and facilitates communication throughout an organization.

Getting by with manual processes or point solutions just ensures that you are spending more time and money developing your product than your competitors and puts you at a serious disadvantage. Additionally, the longer you wait to put a PLM solution in place the more complicated it becomes. You generate legacy information that has to be populated into the system and create process that may not lend itself to being incorporated into a PLM. It may be challenging or seem expensive, but most studies by industry research groups validate the impact PLM has for most companies if done correctly—a good example is the Aberdeen Group study cited in "How Do I Justify an Investment in PLM?"

CHAPTER 9

How Do I Select a Partner?

The Hired Gun:
Learning to Shoot Is the Best Approach

There are a lot of PLM providers and consultants that claim to maximize value and offer audits to help you identify inefficiencies and quantify value during PLM deployment. I have yet to see one of these formulaic solutions that works to the fullest extent of a company's needs, although I am sure there are some strong offerings out there. You tend to get what you pay for in this area. "Hired guns" can be very helpful in cleaning up a sloppy product development process. If they are qualified and knowledgeable they can be very useful. Moreover they can be essential resources for any type of PLM software deployment or process reengineering activity. But they are expensive time wasters if you do not follow some key rules when bringing them in.

We can learn a lot from the lessons of the western movies—*High Plains Drifter* and *The Magnificent*

Seven come to mind. You need to avoid being swayed by a quick draw or a flashy set of Colts and take true measure of your resources. In this chapter we will highlight some best practices in selecting external resources and how to maximize their assistance with value assessment for PLM.

One of the biggest mistakes when bringing on the hired gun is to take them at face value. Back in the old west it was difficult to check references, and reputations were built on a significant amount of mythology, rumor, and just plain gossip. It's not much different today with flashy web sites, cool Powerpoint slides, and slick diagrams. It is easy to be swept up by style over substance, only to find out when the bad guys hit town the hired gun is out the door. When the project gets challenging some hired guns wilt like a cactus flower at high noon—and that's the last thing you want.

Fortunately, in today's business climate you have much more opportunity to research your partners. Usually in a paid engagement, companies will provide a qualified resource, but in the PLM industry qualified resources can be a little hard to find. There are people who know a little about a lot and some that know a lot about a little. Either case can be a problem due to the wide ranging requirements to achieve the full depth and breadth of PLM. To be effective in the role of assessing a company's product development process and devising an approach to continually quantify the value from optimization via a PLM, you need a very experienced resource with good communications skills.

There are several ways to identify these types of resources. The key to selecting an external resource is to interview the actual person that will be working with you and understand their experience level. If at all possible, make time to interview previous clients. Finally, examining sample work from previous engagements is ideal if it is available. If they are reluctant to share it, then you may have uncovered an issue. The main thing is to not take things at face value. Companies can spend a lot of time packaging solutions but deliver very little value. Ultimately, having a handle on who you are working with, and what they are capable of, will increase the odds of success.

The other pitfall of the hired gun is that focusing on that person centers your vision of PLM on the event instead of the process. This means that it can be very easy to focus on the specific activity you are bringing resources to address—a PLM implementation—but not think about the longer term ownership of the application. The problem with this, just like in *High Plains Drifter*, is that once they are gone you are back where you started. You really need to emphasize with your resource that the deliverable is the process not the results of the process. You are looking for assistance in creating an infrastructure to manage the product lifecycle and to continually measure process and quantify value—instead of just having someone come in and install software and review your process and tell you what is wrong with it.

Most consulting companies are not geared for this

approach. They have developed templates and questionnaires to identify issues, but they are not set up to teach you how to do this yourself. The problem is that maintaining a PLM system and measuring value is an ongoing activity not an event. You need to constantly monitor how your product development process is performing and be able to make decisions based on this information. Having a vendor come in periodically and assess things is not ideal and could get pretty expensive. Having a methodology in place to maintain the system and measure how things are working is more likely to yield the results you seek. So in the end you are looking for your hired gun to teach you how to defend yourself versus having them fight your battles for you.

The last potential issue involving the hired gun is company participation. This is somewhat related to the previous issue, but the mindset is that since an external resource is implementing the software, no one from the company needs to be involved. Company participation is key to any type of PLM deployment, and the only way to ensure proper participation is to fully understand who needs to be involved and to what degree. Identifying specific roles is crucial. Most competent consultants should have a good handle on this, but you will need executive sponsorship within the company to make it happen. Executive sponsorship at the proper level, typically a VP or C type, ensures all elements of the company that need to participate will do so and will provide the necessary input to get an accurate picture of status quo process. Without this sponsorship departments may balk if they are not clear on how participating benefits their group or feel like they don't have the time to fully invest in the process. Many times we have seen the hired gun standing alone without the backing of the town and failing. Strong management support is vital to ensure success.

To select a suitable partner, remember these three keys: due diligence, a process instead of an event, and strong company support. If you have these three elements things will more than likely end well with you and the hired gun, riding off into the sunset knowing the job is done well and the company is safe and sound and better off from the experience.

CHAPTER 10

When Should I Move to PLM from PDM?

Home Improvement: Remodel or Replace?

My wife loves HGTV, so we spend a significant amount of time watching shows about home renovation. Sometimes these renovations can be quite extensive, and I often wonder at what point are you better off just finding a new house. I see a similar phenomenon in the PLM space when companies build out or enhance existing applications. Sometimes this is a viable strategy to extend a useful tool to meet additional needs. However, at some point it can go too far. You have to be able to assess when your current product data management (PDM) or PLM system has reached its limits and additional modifications would be ill advised. A PDM system is a basic data vaulting tool typically associated with a CAD system; a PLM, as we've discussed, is much more complex. Adding a gourmet kitchen to 1000-square-foot duplex just doesn't make sense, and trying to use add-ons to turn a PDM system into a PLM system doesn't work either.

The other side of this coin is that if it works for your company there is no need to toss it out and replace it with something more complicated. Like many other things it is a balancing act that requires wisdom and discernment on the part of management. This chapter will analyze what types of modifications for PDM and PLM are reasonable and which ones will tip the scales towards trying to find a new system all together.

Smaller companies and some larger ones typically start out with point solutions that manage data for specific applications or groups inside the company. Eventually they start to recognize the value of PLM and look to expand the footprint of their existing systems by engaging more users throughout the company. But with these new users come new requirements that can tax an application-specific PDM tool beyond its limits. Enterprise PDM from SolidWorks (EPDM) is a good example of an entry level system for data management and workflow. Many smaller companies leverage this product as their first data management tool and, as they grow, seek to expand its capabilities. This approach is sound to a point. Once you start to get into the more complicated aspects of PLM, this strategy starts to fall apart and you reach a point of diminishing returns.

Custom enhancements can enable things like setting up a change control board with voting. EPDM and other tools allow basic routing out of the box, but any formal change control requires some work. This is an area where things have the potential to get out of hand, so change requirements need to be fairly straightforward if you are planning on addressing them with a PDM tool.

Another type of enhancement appropriate for PDM is enabling the system to manage more file types and be accessed from areas outside of CAD. Office2PDM by Razorleaf brings EPDM capability to Microsoft Office applications. This would allow a broader audience to take advantage of EPDM's vaulting technology and allow easy check-in of non-CAD information into the vault. This seems like a useful function and a reasonable addition to a PDM application. Project management is also a natural extension of PDM. 360 Enterprise Software has developed a suite of products that extend EPDM information into the enterprise and link the information to project management tools. Depending upon a company's requirements this could be a reasonable intermediate step to adopting a full-blown PLM solution.

The last potential enhancement for PDM is the ability to export information from PDM directly into an ERP application. This is the point where the differences between PLM and PDM can start to create issues. The problem with ERP is that the type of information it is expecting is fairly far removed from the type of information a PDM application captures. PDM tools are more concerned about files than they are about product structures like the bill of material (BOM).

While the information the ERP needs may exist in some form or fashion in the PDM, getting it out in the right format could be a tall order. Generally, if you are looking to create BOM-type information for an ERP system, you are better served to go ahead and invest in PLM, but there are always exceptions based mainly on how complicated the product structure is and the types of products your company builds.

One of the key differentiators between PDM and PLM is that PDM systems track information in a file structure. This file-oriented structure is why a tool like EPDM is so popular among SolidWorks users. The interface closely resembles the file structure of Windows Explorer. The now-defunct Product Point from PTC offered a similar interface that was also considered very user friendly. PLM tools focus on product structure as opposed to file structure. Physical file attachments are a secondary consideration to the actual structure of the product itself. Sometimes the structure in CAD assemblies mirrors this and sometimes is does not. It really depends on how companies utilize CAD tools and the types of products they make. This is typically driven again by the types of products a company builds, but it can also be affected by the methods used by the design group. There is more value for integration between engineering and manufacturing if the CAD structure is closely tied to how it should appear in configuration management or manufacturing. An enterprise-wide PLM system can encourage this condition.

There is a trend in the industry to source more items for product design that are not created in the CAD system and to include information from electrical design systems as well. Most PDM tools are CAD-specific so they don't do particularly well handling relational data from other CAD tools. One of the reasons PTC discontinued Product Point was its lack of multi-CAD support. If a company's manufacturing BOM is a blend of sourced and made parts coming from multiple systems, then PDM may fall short as a solution and certainly will not be an ideal source for an ERP system.

There are a number of other areas where PLM provides capabilities far beyond PDM and where expanding PDM could be ill-advised. Supply chain management is one of these areas unless the requirement is simply to be able to access CAD information. There are numerous features for supply chain management that include costing, availability, and reporting that can hinder a company's ability to effectively control supplier data. PLM offers a robust set of capabilities for this area that blend functionality from design, quality assurance, and manufacturing to allow companies to optimize their supply chain and meet requirements for cost and time.

Another area where PDM falls short and where customization or enhancements could be challenging are in workflow capabilities. The workflow engines and the logic for configuring them are quite robust in most PLM tools. Recreating anything close to this in PDM tools would be very costly and would ultimately be limited by the architecture of the system. If a company wants to fully automate the change process and implement closed-loop architecture, PLM is the best platform for this due to the built-in process templates and the ability to configure workflows via a user interface. It's also easier to achieve specialized functionality like governance and compliance with both environmental and medical regulations with PLM due to architecture and prebuilt capabilities in PLM.

There is a place for both PDM and PLM, and it is not always a good decision to replace PDM with PLM when just adding a little capability to your existing application will address your immediate needs. There is also a point where enhancements and add-ons will create a unstable environment with limited tools, so executives really need to understand their business objectives thoroughly before making decisions one way or the other. PDM is a good start for a company looking to optimize their product development process and may serve a smaller company's needs for quite some time. It's common to view things in extreme contexts where it is always one way or another, but in the complex situations of most companies, these generalizations don't always apply. Sometimes it is better to fix up what you have rather than bring in an entirely new system. This is the lesson you learn in watching home improvement shows and in the efforts to improve process and product development capabilities.

CHAPTER 11

What About Legacy Data?

What's the Big Deal About Data Migration?

Migration: a word that conjures up fear in the hearts of IT professionals. Migration is associated with hardship and danger when referring to the migration of people. Data migration is no different especially when PLM tools are involved. The most apt description for PLM data migration I have heard was in a meeting with Dell. One of their IT resources described it as "changing out the airplane engine while the plane is still flying without crashing the airplane."

Migration may be an afterthought in a lot of PLM implementation projects but it can dictate the success or failure of PLM and if done improperly it can result in a dramatic loss of altitude ending with an equally dramatic deceleration event as you hit the ground. Zero Wait-State specializes in data migration and we have spent the last decade migrating companies, both large and small, on and off of a variety of PDM and

PLM platforms. This chapter will highlight some of the perils and pitfalls around PLM data migration and best practices that can keep you flying high.

Garbage in Garbage Out

One of the first mistakes companies make when embarking upon a data migration project is assuming their current data structure is worthy of migration. Over the year companies accumulate a large amount of information. Not all of it is critical to the company going forward, and not all of it is in a condition that will make it useful in the new system. This is a good time to take a hard look at your legacy data and determine what really needs to come over into the new system. Older product information may no longer be relevant. Also if data is corrupted or sub-optimal you should think hard before polluting the new system with the data. Some information may not be structured in a way to be useful in a PLM system; this includes having data structured in the same way you want it to manifest itself in the PLM system and having associated information about the file (such as metadata or attributes). When you are moving from a PDM or PLM tool into a new PLM tool, you are able to capture useful metadata like attributes, but many companies migrate from system disks to PLM, which can be much more challenging. Companies need to spend time evaluating current data and its relevance for future product development needs.

2 Gallons of Water in a 1 Gallon Container

There are other issues for companies who are downsizing or as they like to say "rightsizing." Companies that elect to move from a more sophisticated PLM to a more affordable or easier to use system can run into issues if they have more information to move than the new system can accommodate. Examples of this might be moving from Agile Advantage to Arena or from Intralink or PDMLink to Enterprise PDM (EPDM). The source system supports more data types or attributes than the new system can deal with, so you need to take this into account when preparing the information to move over. More appropriately you need to consider this before you make the decision to move. If any of the information is critical, you may be in for an unpleasant surprise. Workflows, change history, file attachments, and access control logic are all things that may not be accommodated when moving from one vendor to another.

What a Tangled Web We Weave

It is not unusual for a company to want to migrate multiple PDMs or data sources to a single environment. While I certainly see the value in doing something like this it adds significant complexity to a data migration effort. Issues like naming conflicts and data integrity can severely hinder a migration involving multiple data sources. Coordination between the sets of information is essential and in-depth analysis before you begin will save major issues from appearing after the migration is complete. Most PLM and PDM systems will not allow nonunique names, so it is best to resolve these issues prior to migrating into the target system. The more consolidation you do prior to migration, the less effort you will expend during and after the migration.

Volume, Volume, Volume

Some of the projects involve massive amounts of data. I am sure you are aware of how much information even a small company can accumulate over time. This creates challenges when it comes to migration. The first challenge is being able to convert all of the information in a timely manner. One of the keys to being able to change out the engine in midair is being able to do it at a time when no new information is being added to the source system. Large data sets can take several days to process so this can be a problem. There are two ways to resolve this issue. One is reducing the amount of data through purging or deletion; the other is to migrate in phases. Phased migration makes sense for most companies because it allows you to break up the activity into manageable pieces. It also allows you to validate and test. The idea is that you take periodic snapshots of the data you are moving and test the data to validate the process. Once the data has been validated, you can do a final migration of just the delta between when you took the snapshot and the most current version of the source data.

History Is Best Forgotten

There is a famous quote from the philosopher George Santayana saying, "Those who cannot remember the past are doomed to repeat it." George obviously has never been involved in data migration projects. History from PDM and PLM systems adds a degree of complexity to a migration that can be prohibitive. Extracting and capturing history from legacy PLM and PDM systems is very challenging. Ensuring that this information gets accurately transmitted into a target system can be even more problematic depending upon the sophistication of the import utilities. Using a "latest only" approach that involves migrating only the latest revision of your data sets dramatically lessens the complexity of the migration, and most companies discover that their history wasn't that useful. If it is absolutely necessary to preserve the history, some companies will elect to keep their legacy system around or create a virtualized server with VMWare or Xen and let people access it on an as-needed basis. For most companies that need is surprisingly less than they expected. This type of migration also addresses some of the issues I mentioned above related to volume and relevance of data.

Be Prepared

The famous Scout motto applies to many things, but especially to PLM data migration. Preparation starts with hardware. It is a big mistake to scrimp on the CPU power and memory for servers for this type of effort, particularly if you are planning to move a lot of data. The other thing to be prepared with is a good set of data analysis and cleansing tools. As we discussed above, understanding the condition of your data and being able to automate some of the cleanup can be very helpful. In most cases, particularly with metadata, there are scripts and applications to detect naming conflicts and validate data transmission on the backend. It is important that you either have these capabilities or work with someone who does. You need to walk through the analysis process and understand the condition of your data so you are not caught off guard by the level of cleanup you will need on the back end of the migration. Do as much clean-up as possible prior to moving information into the target system. It is infinitely easier to clean up information prior to migrating into a new PLM tool.

Test, Validate, Rinse Repeat

It is almost impossible to test too much during a data migration. We recommend a minimum of two test passes, which should allow you to go into a test system and thoroughly sample the data before eventually moving to production. The project should include a well-considered test plan, either developed internally or with a partner's assistance. Good test plans will thoroughly sample the data sets and make sure that the data is functional in the new environment.

Ad hoc testing is better than nothing but there is a good possibility that you could miss something. Generic test plans are better than ad hoc testing, but again, without tailoring something specific for your environment you may find out the hard way that something is missing or wrong. If your company is required to validate per FDA or other regulations you can combine some of this effort into testing and potentially compress your testing cycle. It is best to address testing up front before the project gets too far down the road. The temptation to truncate testing is great, so you must resist.

The End of the Road

Most of these ideas are common sense but are disregarded surprisingly often in data migration efforts. Too many times the migration project is bolted on to a PLM implementation as an afterthought and is completed in the rush to get the new system into production. It is important to recognize that PLM tools are vessels that hold the most important asset a company has: their intellectual property. It is far more important to ensure that information is transferred cleanly and accurately than it is to start up a new PLM system.

Plane crashes are tragic, and obviously anything that happens with migration pales in comparison with the loss of life experienced, but in the context of business nothing can be more crippling than losing product information. This data is the lifeblood of a product development company, so safeguarding it should be one of the company's highest priorities.

CHAPTER 12

What Is the Best Way to Migrate Data?

The Big Bang: Great for Universe Creation Bad for Data Migration

One of the most prevalent and scientifically accepted theories on how the universe was created is the Big Bang theory. According to the ever reliable Wikipedia, "The Big Bang was the event which led to the formation of the universe, according to the prevailing cosmological theory of the universe's early development (known as the Big Bang theory or Big Bang model)."

In the world of data migration a Big Bang would be trying to move all of your data over from one system or environment to another in a single exercise. While Big Bang seemed to work well for creating the universe, I am afraid this approach for data migration is less effective. This chapter will discuss the downside of using a Big Bang approach to data migration and the advantages of incremental migration. This approach is known as the slow-drip method.

How feasible is attempting to migrate data in a compressed timeframe, typically over a weekend? Usually a company identifies their new data management system and sets a go-live date. The objective is to move all of the data from the old system to the new system and then shut down the old system in as short a time as possible. When making a change it is natural to want to get it done as quickly as possible and to not have a lingering transition.

Unfortunately, the nature of data migration does not lend itself to this approach. Most data migration projects involve large amounts of data so there is the practical element of the amount of time it takes to move large amounts of data from one system to another. As we discussed in the previous chapter, this methodology can lead to some serious problems.

In most cases there are two issues that drive a compressed migration schedule. The first is a strong desire to move forward on a new platform and eliminate a legacy system. Often a company faces hardware obsolescence that is undermining the stability and viability of their current system. There is a sense of urgency to move the data onto a new platform before the old system crashes and becomes unusable. You also have situations where a company is dissatisfied with the capabilities of a legacy system and wants to move to a new application to improve user productivity and capability as quickly as possible.

Both of these are strong reasons to migrate quickly, but there are ways to accomplish this without the stress and shortcuts that inevitably occur with rapid data migration. Setting up a virtual server for the legacy data management tool, using software from companies like VMWare, Citrix, or Microsoft, can allow a company to move to newer more reliable hardware and set up backup solutions to protect historical information. This allows a company to better assess over time what information really needs to be in the new system. Productivity gains made by moving to a new system can easily be nullified by importing flawed information into the system. It is better to take time to make sure the data is accurate than to bring it into a new environment and undermine confidence in the new system.

There is a middle ground between a rapid migration and allowing a migration to dictate when you stand up a new system: a slow-drip method. One of our most successful data migrations involved a company that had multiple business units that were releasing products at different times throughout a year. Migrating their data over could potentially disrupt product releases and severely impact the business unit's performance. Given how the company functioned, there would never be a time when all the business units were aligned and could move over to a new platform simultaneously. To address this issue we created tools that would allow each division to migrate over when they were ready.

This allowed each unit to analyze what information they really needed in the new system and to pick the opportune time to shift over. Over the year period all of the business units moved over, and then they shut down the legacy environment.

This approach will not work for every company but moving over distinct product lines or business units is one way to break up a migration into more manageable pieces. Another way to address the issue is to take current projects or versions first and then move over historical information as it is needed. This gives a company more time to analyze the necessity for previous versions of data and also to scrutinize it for data integrity. Moving data over in small increments also allows for cleansing as you go making it more feasible to address missing attributes or broken links or any of the other types of challenges moving CAD data presents.

The desire to move data to a new system in a Big Bang is understandable, but the challenges this type of approach presents can be expensive in both time and money. If it is at all possible extending migration and moving information over in small amounts offers a chance to correct errors and ensure that information is properly structured for a new environment. New technology and methods can help facilitate a gradual migration that will be less traumatic to the organization and ensure a higher likelihood of success with a new system.

Conclusion

When you create a book like *The PLM Primer* it is not a completely selfless endeavor. By creating the reference material in this book and my blog, "The PLM State" I hope to share the philosophy of our company, Zero Wait-State. The name conveys our core value of eliminating the gaps between technology and process. When companies attempt to adopt technology to resolve process issues, the result can be traumatic, and Zero Wait-State was founded to eliminate that trauma by helping companies make the right choices and take effective ownership of technology adoption.

We offer companies the benefit of our experience in marrying these technologies to their business process. We combine our technical know-how with real-world industry experience to provide our clients with pragmatic guidance on how best to adopt PLM. Specifically, we can provide up front input that can help your company avoid the most common mistakes in technology adoption. We can supplement your organization with our bandwidth to make sure the planning necessary for success happens. Then we can walk you through the steps for a seamless implementation and transition from legacy systems.

If your company is struggling to get products to market or with quality or cost issues, PLM can be a valuable solution. If your company is in its early stages, PLM is a foundational ingredient for your future success. Please come to our website to learn more, and feel free to contact me personally if you would like discuss PLM. Zero Wait-State's website is **www.zerowait-state.com**.

Acknowledgements

Abook like this does not get written by just one person even though my name is listed as the author. Technically I wrote the entire book but it could not have happened without the assistance of numerous co-workers, clients and partners that I have encountered throughout my career. A number of people have been very encouraging and helpful since I started my blog, "The PLM State" three years ago. I would specifically like to thank Shane Goodwin and John Kelley from Oracle for their positive feedback and technical input. I would also like to thank Alex Dye from Masterlock for his enthusiastic support and great ideas for future articles. Finally, I would like to extend gratitude to my partner David Stewart and co-workers Steve Ammann and Paul Peck whose knowledge and experience with PLM and process improvement have provided me with great inspiration for many of the articles I have written.

I also want to thank my son Adam. His appreciation of literature and modern movies keeps me in touch with the value of entertainment which I feel is a critical element in effectively communicating information like PLM. Technical subjects can be very tough to digest and he has taught me how to communicate in a manner that I believe reaches a broader audience. I look forward to one day reading one of his books.

CPSIA information can be obtained at www.ICGtesting.com
Printed in the USA
LVOW09s1759010215

425238LV00026B/1103/P

9 780988 618305